Match Up!

AF694226

Find the tally marks that show the number.

1 9	2 3	3 16	4 8
5 10	6 12	7 5	8 11
9 15	10 23	11 20	12 14

Answer Box

A	B	C	D	E	F
𝍸 𝍸 𝍸 𝍸 |||	𝍸 |||	𝍸 𝍸 ||	|||	𝍸 𝍸 𝍸 𝍸	𝍸 𝍸 𝍸 |
G	**H**	**I**	**J**	**K**	**L**
𝍸 ||||	𝍸	𝍸 𝍸 |	𝍸 𝍸 ||||	𝍸 𝍸 𝍸	𝍸 𝍸

Find the tally marks that show the number.

1 4	2 10	3 13	4 21
5 8	6 15	7 18	8 16
9 7	10 6	11 20	12 12

Answer Box

A	B	C	D	E	F
𝍸 ||	𝍸 𝍸 |||	𝍸 𝍸	𝍸 𝍸 𝍸	||||	𝍸 𝍸 𝍸 𝍸 |
G	**H**	**I**	**J**	**K**	**L**
𝍸 𝍸 𝍸 𝍸	𝍸 |||	𝍸 𝍸 𝍸 |	𝍸 𝍸 ||	𝍸 |	𝍸 𝍸 𝍸 |||

Objective: Identify the tally that represents a number.

Trip Tally

Use the table to answer the question.

How many students voted for:

1. the hiking trail?
2. the farm?
3. the aquarium?
4. the zoo?
5. the historic house or the aquarium?
6. the hiking trail or the science museum?

Which trip received:

7. 20 votes?
8. the most votes?
9. fewer votes, the farm or the hiking trail?
10. the same number of votes as the hiking trail?
11. the least votes?
12. more votes, the historic house or the farm?

Answer Box

A	B	C	D	E	F
35	18	9	Hiking trail	21	15
G	**H**	**I**	**J**	**K**	**L**
Farm	Zoo	24	Aquarium	Science museum	Historic house

 Objective: Interpret and compare data given in a tally table.

Recycle and Tally!

Use the table to answer the question.

Recycling Tally	
Newspapers	卌 卌
Magazines	卌 卌 III
Plastic bottles	卌 卌 卌 卌
Green glass bottles	卌 III
Clear glass bottles	卌 卌 卌 I
Cans	卌 卌 卌 卌 II

How many:

1. cans did the family recycle?
2. glass bottles did they recycle?
3. magazines and newspapers did they recycle in all?
4. bottles did the family recycle in all?
5. items that are not bottles did the family recycle?
6. more plastic bottles than green glass bottles did they recycle?
7. Which item did the family recycle the least?
8. Which item did they recycle the most?
9. Did the family recycle more magazines or newspapers?
10. Did they recycle more plastic bottles or green glass bottles?
11. Did they recycle fewer cans or clear glass bottles?
12. Did they recycle fewer clear glass bottles or newspapers?

Answer Box

A	B	C	D	E	F
22	44	23	Magazines	24	12
G	**H**	**I**	**J**	**K**	**L**
Green glass bottle	Plastic bottles	45	Clear glass bottles	Newspapers	Cans

Objective: Interpret and compare data given in a tally table.

Problem Solving: Using a Timeline

1804	1839	1881	1913	1937
First steam locomotive built	Pedal bicycle invented	First electric railroad opens	Model T car is produced	Jet engine designed

[Source: Essential Facts 1996, Dorling Kindersley Ltd., London]

Use the timeline to solve the problem.

1. When was the pedal bicycle invented?
2. When was the first non-stop flight around the world?
3. When did an astronaut first land on the moon?
4. Was the Model T produced before or after the first steam locomotive?
5. You can use this timeline to find out how long ago the first person went into space. True or false?
6. How many years are there between the first and last events shown on this timeline?
7. Did an astronaut travel in space before or after the non-stop flight around the world?
8. How many years after the first electric railroad opened did an electric train travel more than 300 mph?

1952	1961	1969	1986	1990
World's first jet airliner produced	First person in space	First astronauts on the moon	First non-stop flight around the world	Electric train travels 320 mph

9 Franklin's mother was born 3 years before the first astronauts landed on the moon. In what year was she born?

10 You can use this timeline to find how long after a jet engine was designed that a rocket was designed. True or false?

11 How many years are there between the time the jet engine was designed and the time the first jet airliner was produced?

12 Mr. Ford bought a car that was made 83 years after the Model T was first produced. In what year was Mr. Ford's car made?

Answer Box

A	B	C	D	E	F
1969	15	Before	False	1996	1986
G	**H**	**I**	**J**	**K**	**L**
True	1966	After	1839	186	109

Objective: Solve a problem by using information given in a simple timeline.

Let's Run!

Use the graph to answer the question.

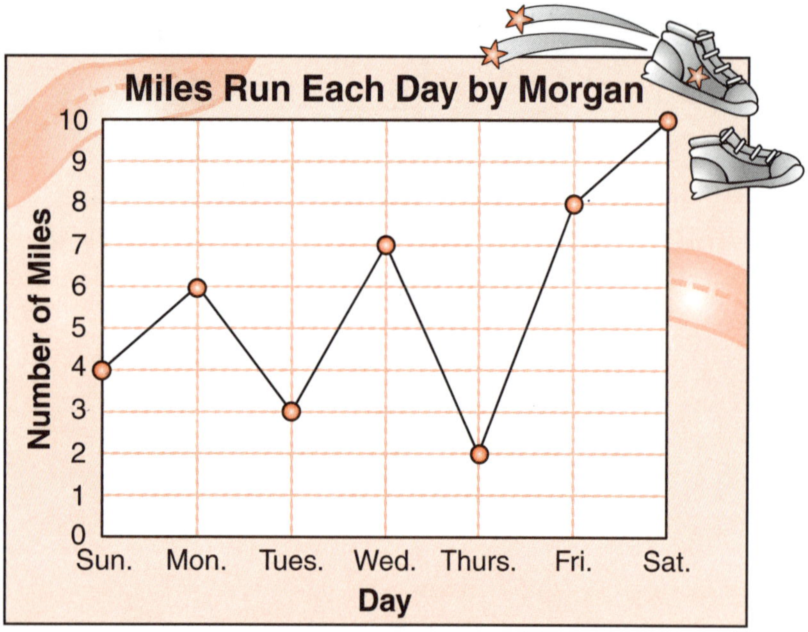

On which day did Morgan run:

1. 3 miles?
2. 8 miles?
3. more than 8 miles?
4. less than 3 miles?
5. more miles, Thursday or Monday?
6. more miles, Sunday or Tuesday?

How many:

7. miles did Morgan run on Monday?
8. miles did she run on the weekend?
9. miles is the greatest number Morgan ran on one day?
10. more miles did Morgan run on Friday than on Tuesday?
11. days did Morgan run more than 5 miles?
12. miles is the least number Morgan ran on one day?

Answer Box

A	B	C	D	E	F
Friday	2	5	Thursday	10	Monday
G	**H**	**I**	**J**	**K**	**L**
14	Sunday	Tuesday	6	4	Saturday

Objective: Interpret and compare data displayed in a line graph.

Please Pass the Popcorn!

Use the graph to answer the question.

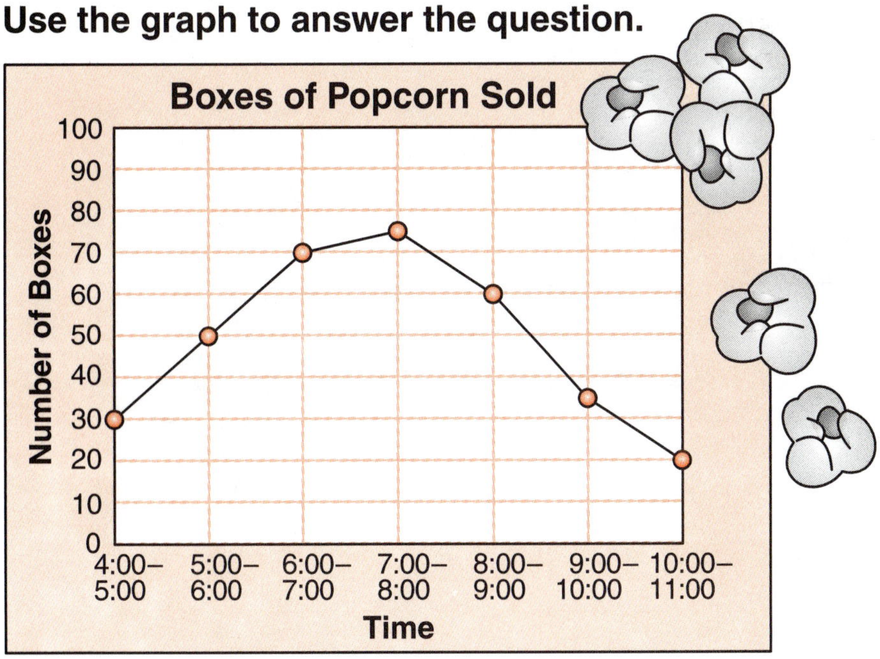

How many boxes were sold:

1. from 5:00 to 6:00?
2. from 7:00 to 8:00?
3. from 9:00 to 10:00?
4. from 6:00 to 7:00?
5. after 9:00?
6. before 6:00?

During which times were:

7. 60 boxes sold?
8. the fewest boxes sold?
9. the greatest number of boxes sold?
10. more boxes sold, 10:00 to 11:00 or 4:00 to 5:00?
11. more boxes sold, 5:00 to 6:00 or 9:00 to 10:00?
12. fewer boxes sold, 6:00 to 7:00 or 7:00 to 8:00?

Answer Box

A	B	C	D	E	F
8:00–9:00	50	10:00–11:00	6:00–7:00	35	5:00–6:00
G	**H**	**I**	**J**	**K**	**L**
70	4:00–5:00	80	75	55	7:00–8:00

Objective: Interpret and compare data displayed in a line graph.

Coin Collecting

Use the graph to answer the question.

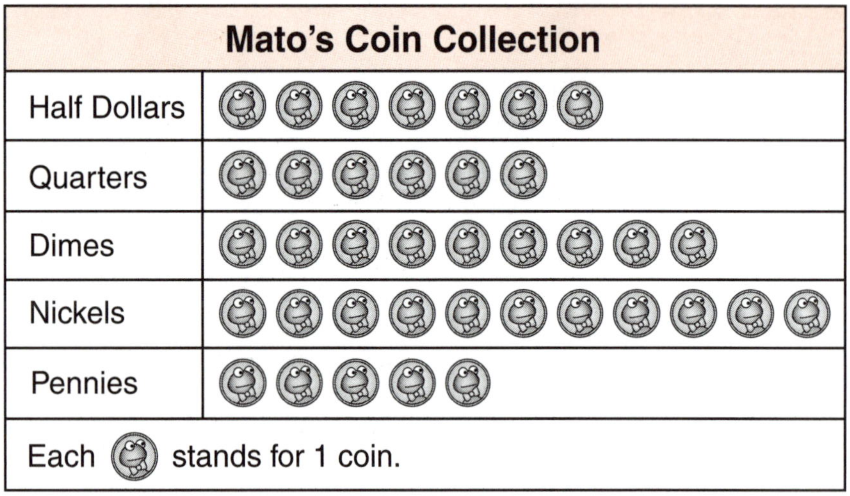

Does Mato have:

1. more nickels or dimes?
2. more half dollars or dimes?
3. more quarters or half dollars?
4. more quarters or pennies?

How many:

5. dimes does Mato have?
6. pennies does Mato have?
7. quarters does he have?
8. half dollars does he have?
9. more nickels than half dollars does Mato have?
10. fewer quarters than dimes does Mato have?
11. coins does Mato have that are worth at least 25¢?
12. coins does Mato have that are worth less than 10¢?

Answer Box

A	B	C	D	E	F
4	9	7	16	3	Quarters
G	**H**	**I**	**J**	**K**	**L**
13	Half dollars	5	Dimes	Nickels	6

 Objective: Interpret and compare data given in a pictograph.

Let's Go Out to Eat!

Use the graph to answer the question.

What meal was ordered by:

1. 18 people?
2. 10 people?
3. the fewest people?
4. the most people?
5. more people, soup and salad or beef stew?
6. more people, chicken cutlet or spaghetti?

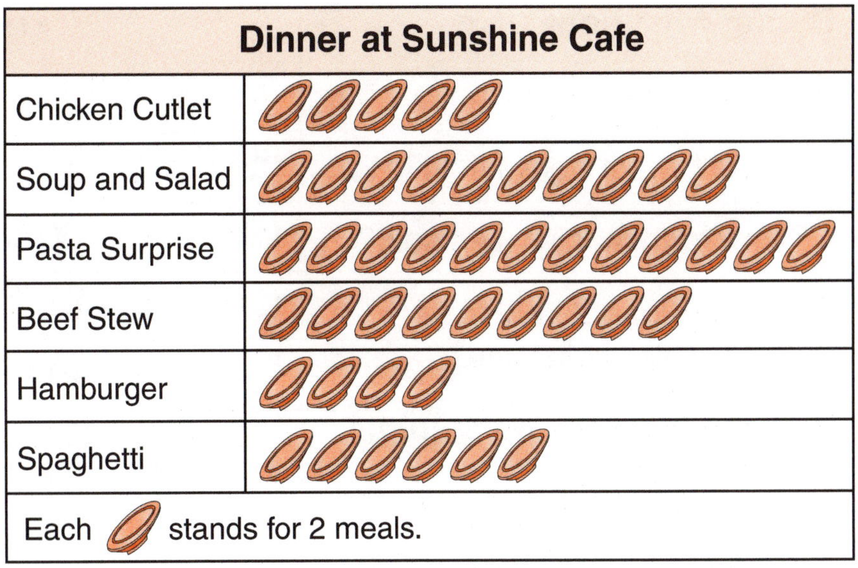

How many:

7. people ordered spaghetti?
8. people ordered soup and salad?
9. people ordered hamburger or pasta surprise?
10. people ordered spaghetti or beef stew?
11. more people ordered soup and salad than ordered beef stew?
12. more people ordered spaghetti than ordered hamburger?

Answer Box

A	B	C	D	E	F
20	4	2	Soup and salad	12	30
G	**H**	**I**	**J**	**K**	**L**
Spaghetti	Chicken cutlet	32	Hamburger	Beef stew	Pasta surprise

Objective: Interpret and compare data given in a pictograph.

Book Sale

Use the graph to answer the question.

During what hour were:

1. 30 books sold?
2. the fewest books sold?
3. more than 45 books sold?
4. more than 20 but fewer than 30 books sold?

How many books were sold:

5. between 10:00 and 11:00?
6. between 11:00 and 1:00?
7. after 12:00?
8. in all?
9. from 9:00 to 11:00?
10. from 10:00 to 1:00?
11. if there were only 7 book symbols on the graph?
12. if there were only 9 book symbols on the graph?

Answer Box

A	B	C	D	E	F
75	85	20	35	45	140
G	**H**	**I**	**J**	**K**	**L**
9:00–10:00	1:00–2:00	50	12:00–1:00	11:00–12:00	65

Objective: Interpret and compare data displayed in a pictograph.

Vehicle Volume

Use the bar graph to answer the question.

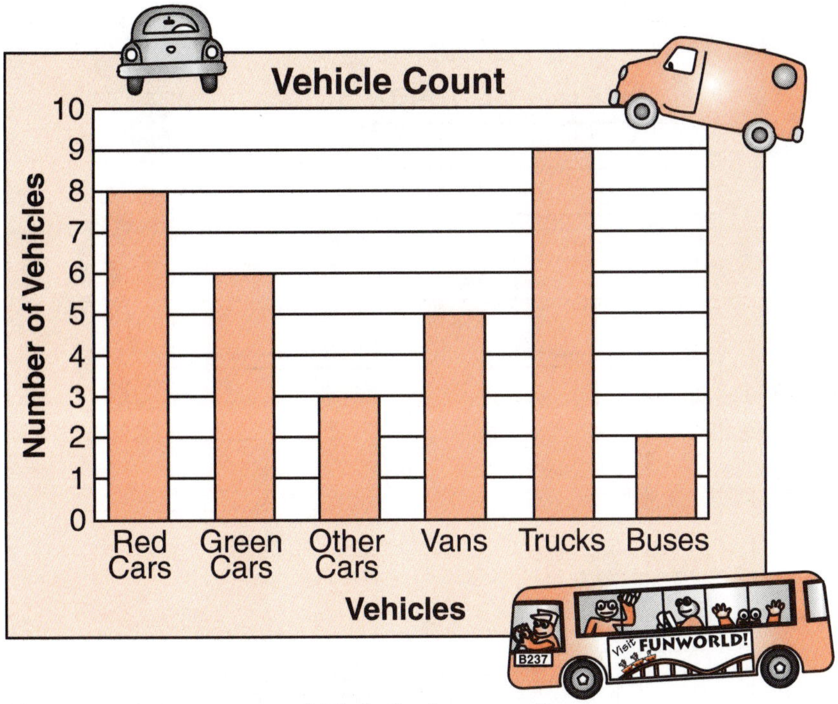

How many passed Eric's house?

1. green cars
2. buses
3. cars that were not red or green
4. vehicles that were not cars
5. cars in all
6. trucks
7. trucks and buses
8. buses and vans

Did Eric count more:

9. vans or buses?
10. green cars or vans?
11. red cars or trucks?
12. red cars or green cars?

Answer Box

A	B	C	D	E	F
17	Trucks	6	16	7	Red cars
G	**H**	**I**	**J**	**K**	**L**
Vans	3	11	Green cars	9	2

Objective: Interpret and compare data given in a vertical bar graph.

School Supplies

Use the graph to answer the question.

Which item sold:

1. the most?
2. more, pens or markers?
3. fewer, pencils or markers?
4. the least?

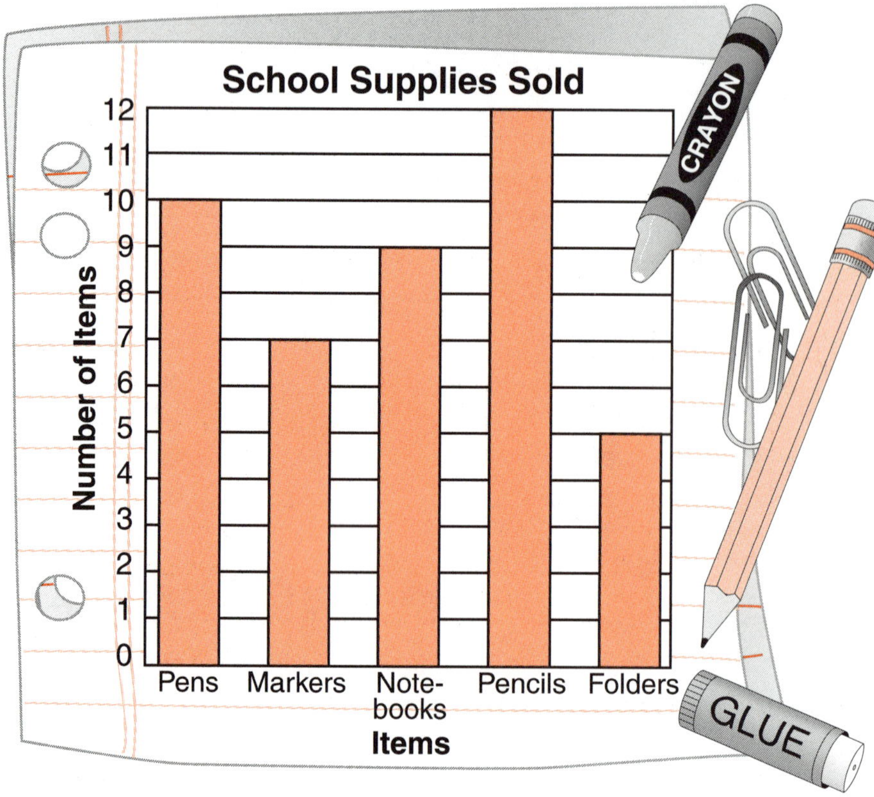

How many:

5. markers were sold?
6. notebooks were sold?
7. pens and markers were sold?
8. notebooks and folders were sold?
9. more notebooks than folders were sold?
10. fewer pens than pencils were sold?
11. items that a student can write with were sold?
12. more pens and pencils than markers were sold?

Answer Box

A	B	C	D	E	F
14	15	2	17	29	7
G	**H**	**I**	**J**	**K**	**L**
9	Pens	Folders	Pencils	Markers	4

 Objective: Interpret and compare data given in a vertical bar graph.

Getting the Jump on Graphs

Use the graph to answer the question.

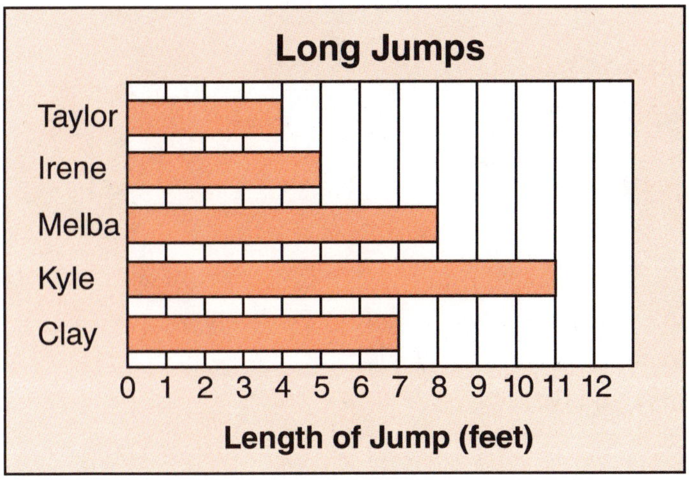

Who jumped:

1. 7 feet?
2. farther, Irene or Taylor?
3. the second longest jump?
4. a shorter distance, Clay or Taylor?

How many:

5. feet did Melba jump?
6. feet did Irene jump?
7. feet long was the shortest jump?
8. feet long was the longest jump?
9. students jumped more than 7 feet?
10. students jumped less than 5 feet?
11. feet farther did Kyle jump than Melba?
12. feet longer was the longest jump than the shortest jump?

Answer Box

A	B	C	D	E	F
Taylor	Clay	5	Irene	2	Melba
G	**H**	**I**	**J**	**K**	**L**
1	4	7	11	8	3

Objective: Interpret and compare data given in a horizontal bar graph.

Problem Solving: Analyzing Data

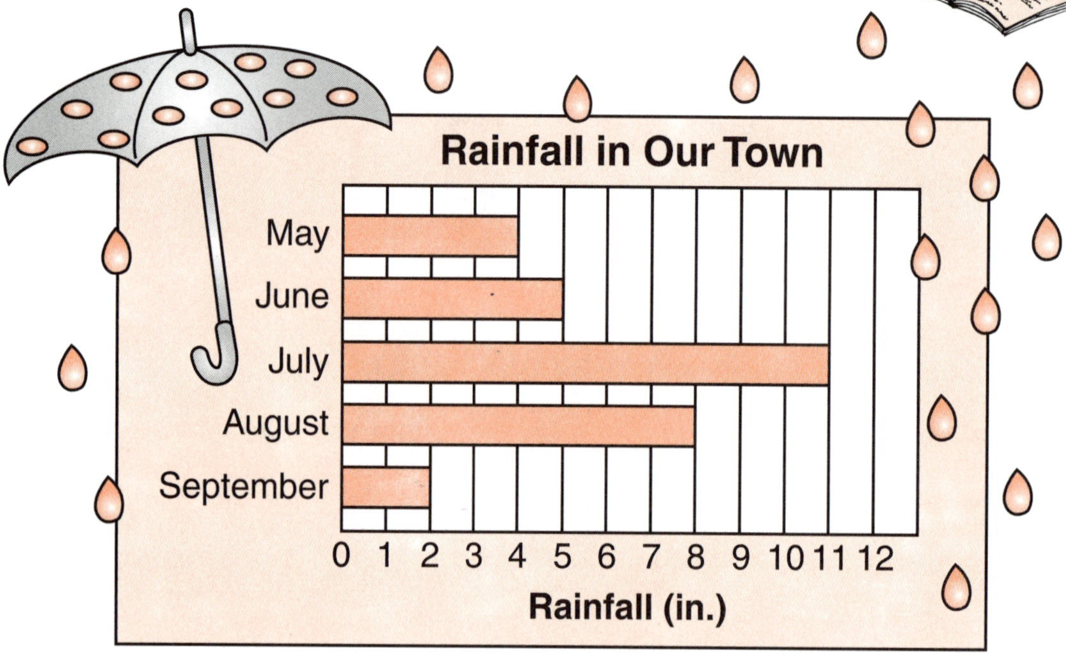

Use the graph to solve the problem.

1. The Norris family wants to take a vacation during May, June, July, August, or September. They do not want rainy days. Which month should they choose?
2. You can use this graph to compare the rainfall during different months. Yes or no?
3. How much more rain fell in June than in May?
4. It was drier in the middle of the summer than at the beginning or at the end. True or false?
5. Keisha wants to visit her cousin between May and September. She and her cousin like rainy days. Which month should Keisha plan her trip?

6. How much less rain fell in August than in July?
7. You can use this graph to compare the rainfall this year to the rainfall last year. Yes or no?
8. How much rain fell in June, July, and August?
9. As the summer started, the rainfall increased. Then it decreased. True or false?
10. How much rain fell altogether in the two wettest months?
11. How much rain fell altogether in the two driest months?
12. How much more rain fell in the wettest month than in the driest month?

Answer Box

A	B	C	D	E	F
True	1 in.	9 in.	24 in.	False	Sept.
G	**H**	**I**	**J**	**K**	**L**
July	6 in.	3 in.	Yes	19 in.	No

Objective: Solve a problem by drawing conclusions from data given in a bar graph.

Tickets for Sale!

Use the graph to answer the question.

Which event costs:

1. $12 for a ticket?
2. $8 for a ticket?
3. the most?
4. the least?
5. less, a movie or a concert?
6. more, a flower show or a car show?

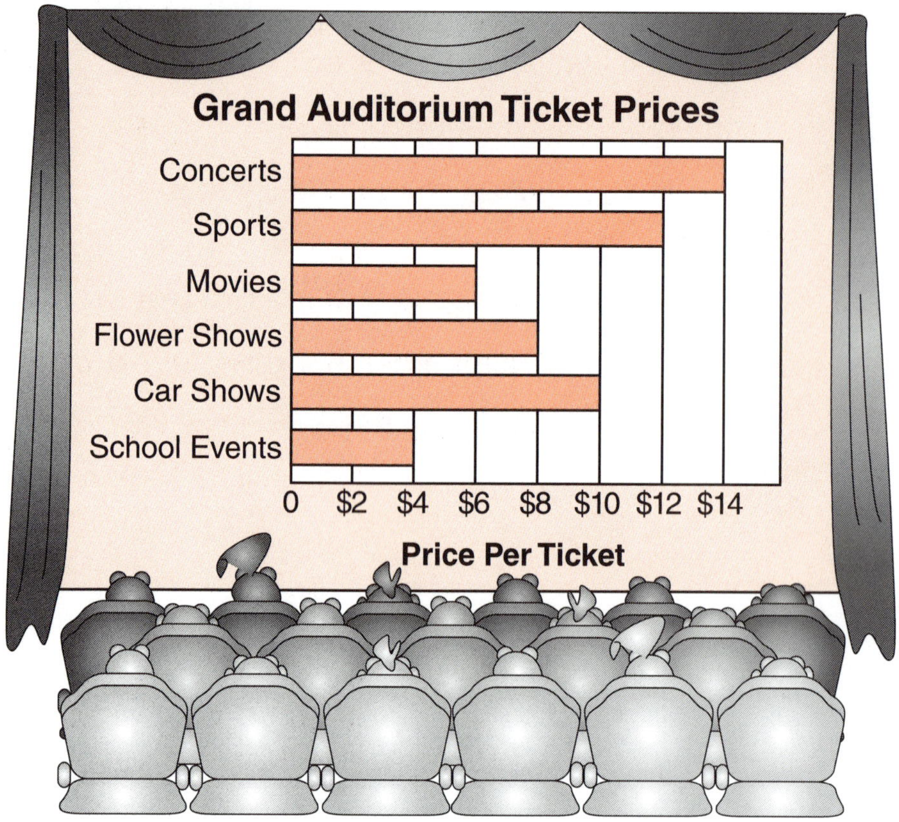

What is the cost of:

7. a movie ticket?
8. two tickets to a sports event?
9. two concert tickets?
10. four tickets to the flower show?
11. three tickets for a car show?
12. four tickets to a school event?

Answer Box

A $16	B $24	C Concert	D $28	E $6	F $32
G Car show	H Flower show	I $30	J Movie	K School event	L Sports

 Objective: Interpret and compare data given in a horizontal bar graph.

What Are the Outcomes?

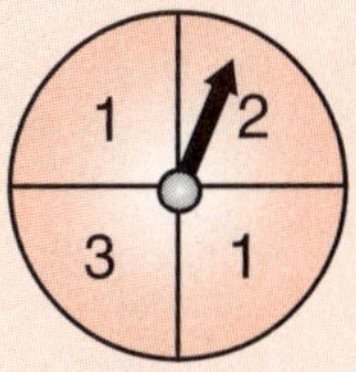

This spinner has 4 possible **outcomes**:

1, 1, 2, 3.

Find the list of outcomes for the spinner.

1. 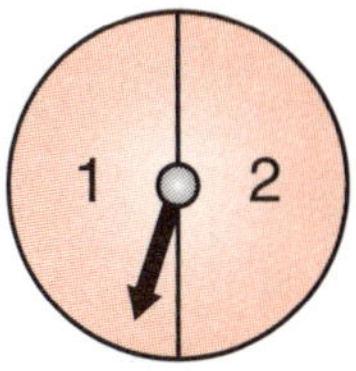

2. 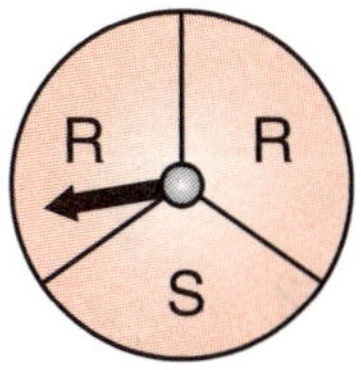

3. 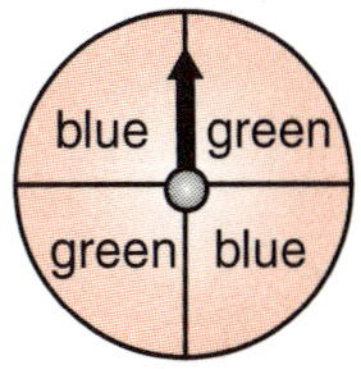

4. 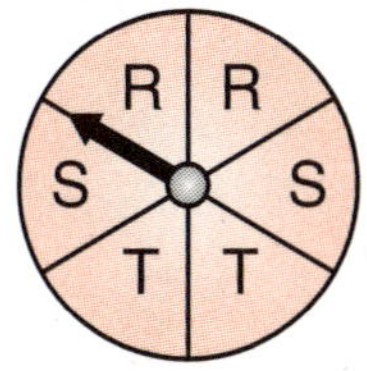

5. 

6. 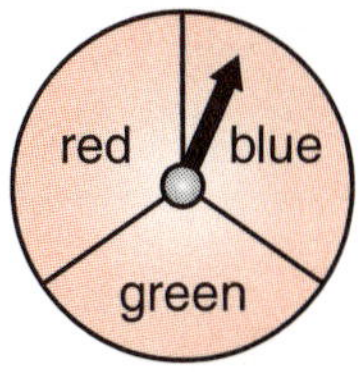

7. 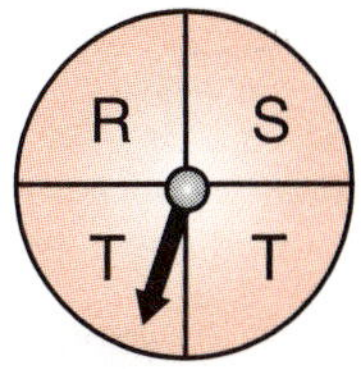

8. 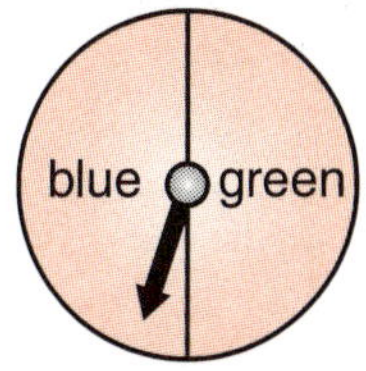

9. 

10. 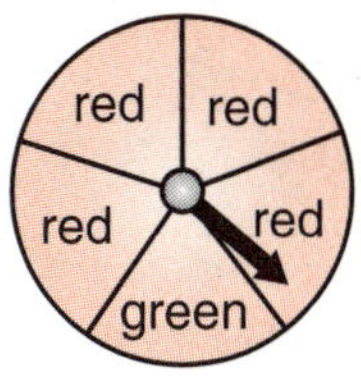

11. 

12. 

Answer Box

A	B	C	D	E	F
red, red, red, red, green	R, S, T, T	1, 2, 3	1, 1, 2, 2	blue, green	R, R, R, R, S, S, S, S
G	**H**	**I**	**J**	**K**	**L**
R, R, S, S, T, T	1, 2	1, 1, 2, 2, 3, 3	blue, blue, green, green	red, blue, green	R, R, S

Objective: Complete a list of all the outcomes generated by spinning a spinner.

Spinning for Outcomes

Find all possible outcomes for the spinner.

1 

2 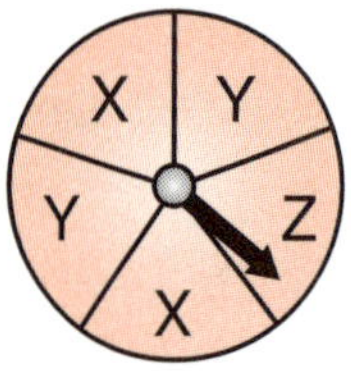

3 

4 

5 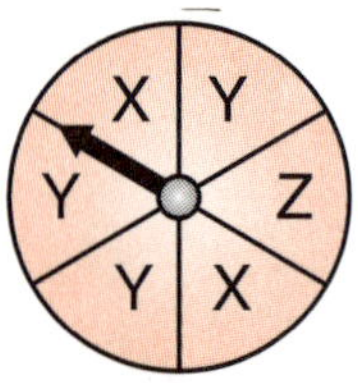

6 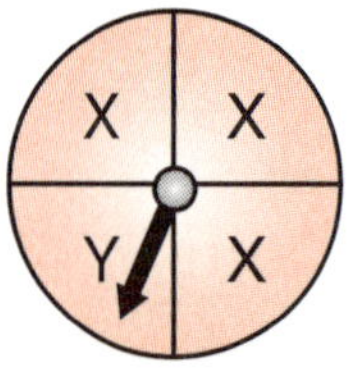

7 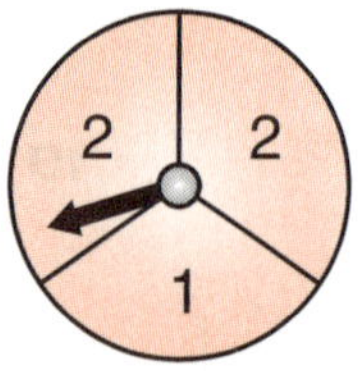

8 

9 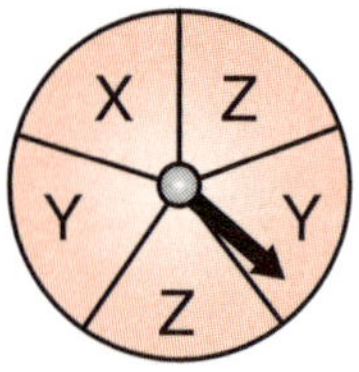

10 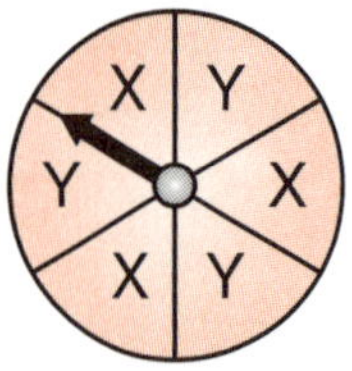

11 

12 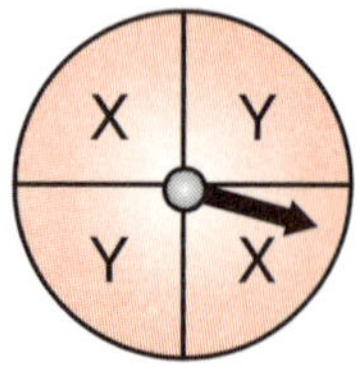

Answer Box

A	B	C	D	E	F
X, X, Y, Y, Y, Z	1, 2, 3, 3	1, 2, 2	X, X, X, Y, Y, Y	X, X, Y, Y	1, 1, 1, 2, 3
G	**H**	**I**	**J**	**K**	**L**
1, 1, 2, 2, 3	X, Y, Y, Z, Z	1, 1, 2, 3	1, 1, 2	X, X, X, Y	X, X, Y, Y, Z

Objective: Complete a list of all the outcomes generated by spinning a spinner.

Card Outcomes

A B C
A B

When picking a card from this group of cards, without looking, there are **5** possible outcomes: A, A, B, B, C.

Find all possible outcomes for the group of cards.

1. 2 4 6
2. X Y X
3. M N N O M M
4. X Y
5. 4 2 2 6
6. M O M O N M N M
7. W Y X X Z W
8. M N O
9. 2 4 6 2 6
10. W X Y Z
11. M N O M N
12. 2 2 6 4 2 6

Answer Box

A	B	C	D	E	F
M, M, M, M, N, N, O, O	X, X, Y	W, X, Y, Z	W, W, X, X, Y, Z	M, M, M, N, N, O	2, 2, 4, 6, 6
G	**H**	**I**	**J**	**K**	**L**
2, 2, 2, 4, 6, 6	M, N, O	X, Y	2, 4, 6	2, 2, 4, 6	M, M, N, N, O

Objective: Complete a list of all the outcomes generated by picking a card from a group of cards.

More Outcomes

Find all possible outcomes for the group of cards.

1. 1 3 5 7
2. P Q R P R
3. red blue red blue
4. P Q R S
5. P Q R S Q P R
6. blue red blue
7. P Q R R S P
8. 1 3 5 7 3 1 5
9. red blue red
10. 1 3 3 5
11. P Q R R Q P
12. 1 3 5 1 7

Answer Box

A	B	C	D	E	F
P, Q, R, S	1, 3, 3, 5	blue, blue, red	1, 1, 3, 3, 5, 5, 7	red, red, blue, blue	P, P, Q, Q, R, R
G	**H**	**I**	**J**	**K**	**L**
P, P, Q, R, R, S	1, 3, 5, 7	1, 1, 3, 5, 7	P, P, Q, R, R	P, P, Q, Q, R, R, S	red, red, blue

Objective: Complete a list of all the outcomes generated by picking a card from a group of cards.

What's the Probability?

Example

Use the spinner to answer the question.

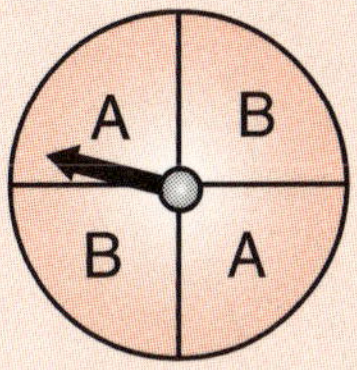

What is the probability of spinning an A?

$$\frac{\text{number of ways to spin an A}}{\text{number of possible outcomes}} \begin{matrix} \rightarrow 2 \\ \rightarrow 4 \end{matrix}$$

So, the probability of spinning an A is $\frac{2}{4}$.

Use the spinner at the right to answer the question.

What is the probability of:

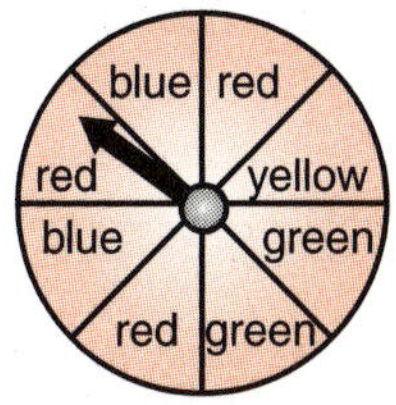

1. spinning yellow?
2. spinning purple?
3. spinning red?
4. spinning blue?
5. What color are you most likely to spin?
6. What color are you least likely to spin?

Use the spinner at the right to answer the question.

What is the probability of:

7. spinning 1?
8. spinning 2?
9. spinning 3?
10. spinning 4?
11. What number are you most likely to spin?
12. What number are you least likely to spin?

Answer Box

A	B	C	D	E	F
$\frac{3}{10}$	$\frac{3}{8}$	$\frac{2}{10}$	yellow	$\frac{1}{8}$	$\frac{2}{8}$
G	**H**	**I**	**J**	**K**	**L**
1	red	$\frac{0}{8}$	4	$\frac{1}{10}$	$\frac{4}{10}$

Objective: Use a fraction to describe the probability of an outcome when spinning a spinner.

How Likely Is It?

Use the spinner at the right to find the answer.

1. You are most likely to spin a ■.
2. You are least likely to spin a ■.
3. The probability of spinning a circle is ■.
4. The probability of spinning a square is ■.
5. The probability of spinning a pentagon is ■.
6. The probability of spinning a triangle is ■.

Use the spinner at the right to find the answer.

7. The probability of spinning an X is ■.
8. The probability of spinning a Z is ■.
9. The probability of spinning a Y is ■.
10. The probability of spinning a W is ■.
11. You are least likely to spin a ■.
12. You are most likely to spin an ■.

Answer Box

A	B	C	D	E	F
X	Triangle	$\frac{0}{8}$	Z	$\frac{3}{8}$	$\frac{0}{10}$
G	**H**	**I**	**J**	**K**	**L**
$\frac{1}{8}$	$\frac{2}{10}$	$\frac{4}{8}$	$\frac{3}{10}$	$\frac{5}{10}$	Circle

Objective: Use a fraction to describe the probability of an outcome when spinning a spinner.

Spin to Win!

Use the spinner at the right to answer the question.

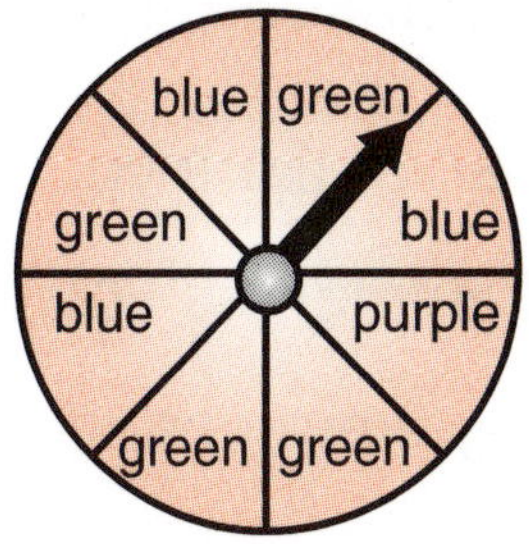

1. What color are you most likely to spin?
2. What color are you least likely to spin?
3. What is the probability of spinning yellow?
4. What is the probability of spinning blue?
5. What is the probability of spinning green?
6. What is the probability of spinning purple?

Use the spinner at the right to answer the question.

7. What number are you most likely to spin?
8. What number are you least likely to spin?
9. What is the probability of spinning 10?
10. What is the probability of spinning 20?
11. What is the probability of spinning 40?
12. Spinning what number has a probability of $\frac{4}{10}$?

Answer Box

A	B	C	D	E	F
20	green	$\frac{4}{8}$	$\frac{1}{10}$	30	purple
G	**H**	**I**	**J**	**K**	**L**
$\frac{3}{8}$	$\frac{5}{10}$	$\frac{0}{8}$	$\frac{0}{10}$	$\frac{1}{8}$	10

Objective: Use a fraction to describe the probability of an outcome when spinning a spinner.

Probability Pick

Suppose the animal cards at the right are placed face down on a table and you pick one.

Use the animal cards to answer the question.

1. What animal are you most likely to pick?
2. What animal are you least likely to pick?
3. What is the probability of picking turtle?
4. What is the probability of picking a pig?
5. What is the probability of picking a horse?
6. What is the probability of picking a dog?

Suppose the number cards at the right are placed face down on a table and you pick one.

Use the number cards to answer the question.

14	12	18	16
16	12	14	12
	12	14	

7. What number are you most likely to pick?
8. What number are you least likely to pick?
9. What is the probability of picking 14?
10. What is the probability of picking 12?
11. What is the probability of picking 18?
12. What is the probability of picking 16?

Answer Box

A	B	C	D	E	F
$\frac{1}{10}$	$\frac{3}{6}$	$\frac{4}{10}$	12	18	$\frac{2}{10}$
G	**H**	**I**	**J**	**K**	**L**
$\frac{3}{10}$	$\frac{2}{6}$	Horse	$\frac{1}{6}$	Pig	$\frac{0}{6}$

Objective: Use a fraction to describe the probability of an outcome when picking a card from a group of cards.

Probability Draw

Suppose the letter cards at the right are placed face down on a table and you pick one.

Use the letter cards to find the answer.

1. You are most likely to pick an ■.
2. You are least likely to pick an ■.
3. The probability of picking a Q is ■.
4. The probability of picking an S is ■.
5. The probability of picking an M is ■.
6. The probability of picking an R is ■.

Suppose the shape cards at the right are placed face down on a table and you pick one.

Use the shape cards to find the answer.

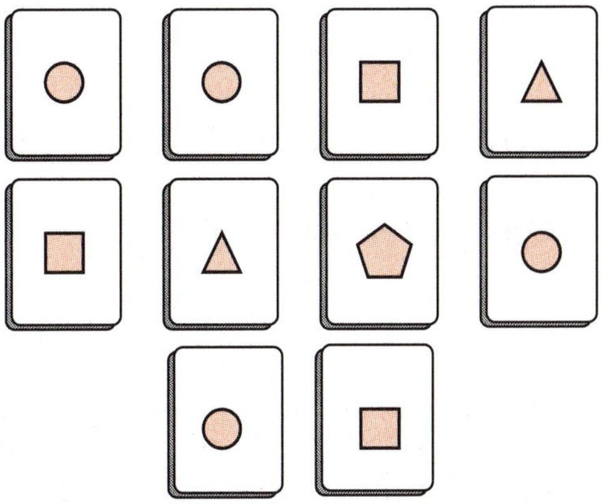

7. The probability of picking a pentagon is ■.
8. The probability of picking a circle is ■.
9. The probability of picking a triangle is ■.
10. The probability of picking a square is ■.
11. You are most likely to pick a ■.
12. You are least likely to pick a ■.

Answer Box

A	B	C	D	E	F
M	$\frac{4}{8}$	$\frac{3}{8}$	$\frac{3}{10}$	$\frac{2}{10}$	$\frac{0}{8}$
G	**H**	**I**	**J**	**K**	**L**
Pentagon	$\frac{4}{10}$	$\frac{1}{10}$	R	$\frac{1}{8}$	Circle

Objective: Use a fraction to describe the probability of an outcome when picking a card from a group of cards.

Color Cards

Suppose the color cards at the right are placed face down on a table and you pick one.

Use the color cards to answer the question.

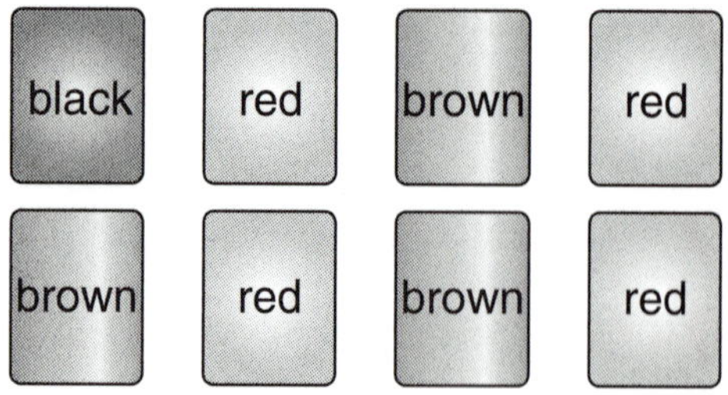

1. What color are you most likely to pick?
2. What color are you least likely to pick?
3. What is the probability of picking red?
4. What is the probability of picking blue?
5. What is the probability of picking brown?
6. What is the probability of picking black?

Suppose the shaded cards at the right are placed face down on a table and you pick one.

Use the shaded cards to answer the question.

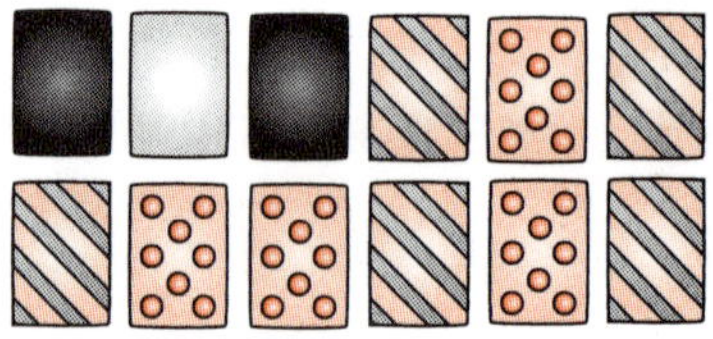

7. What is the probability of picking 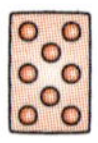 ?
8. What is the probability of picking  ?
9. What is the probability of picking  ?
10. What is the probability of picking  ?
11. What card are you most likely to pick?
12. What card are you least likely to pick?

Answer Box

A	B	C	D	E	F
$\frac{0}{8}$	red	$\frac{1}{8}$	black	$\frac{4}{8}$	$\frac{5}{12}$
G	**H**	**I**	**J**	**K**	**L**
$\frac{2}{12}$	$\frac{4}{12}$		$\frac{1}{12}$		$\frac{3}{8}$

Objective: Use a fraction to describe the probability of getting an outcome when picking a card from a group of cards.